THE END OF BIG BANG COSMOLOGY

Bijective Physics

A. S. Šorli
Bijective Physics Institute
Slovenia

The cover photo source is
European Southern Observatory

CONTENTS

1. The End Big Bang cosmology

Physics needs re-examination and rejuvenation. In this book, we will do this going together step by step. We will re-examine physics only on the basis of experimental data. This will not be a journey into the labyrinths of theoretical speculations. If you like exotic physics, this book is not for you. You better read Michio Kaku, Carlo Rovelli, Brian Green or Lee Smolin. This book is for people who are ready to encounter the truth about today's physics. For this, you need courage and independent free intelligence.

Universe is Infinite in Volume

Back in 2014, NASA has measured that universal space has Euclidean space. They measure angles of the triangle made by three stellar objects and the sum was 180 degrees. This means that universal space has the Euclidean shape and infinite volume. The amount of energy in the form of space is infinite, the amount of energy in the form of matter is infinite.

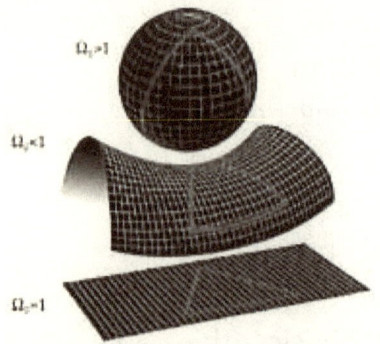

$\Omega_t > 1$

$\Omega_s < 1$

$\Omega_0 = 1$

Universal space is flat

E matter = ∞, E space = ∞, E universe = ∞

That universe has an infinite amount of energy is the fact-based on NASA measurements. We will take this fact as one of our arguments against the existent Big Bang model.

Universe is Time-Invariant
Time is what we measure with the clocks. With the clocks, we measure the duration of material changes, i.e. motion in space. On the basis of our elementary perception, we can conclude that changes run in space only and time is their duration. In entire

physics, we do not have a single experiment which would prove that changes run in time. We will drop this ever-proved idea and accept the fact that time is the duration of changes in space. This means that the universe runs in space only (not in time) and that cosmological principle is time-invariant. The universe as we see it today does not depend on time.

Material changes in the universe are running into space and are irreversible. When change X+1 enters the existence, the change X is not in existence anymore. When change X+2 enters into existence, the change X+1 is not in existence anymore. Time is the numerical sequential order of material changes running in space. Every elapsed time **t** is the sum of Planck times. A Planck time is the fundamental unit of the numerical order of changes.

$t = tp1 + tp2 + ...,+ tpN$

Maybe this is a bit hard to mentally digest, but elementary perception and experimental data are supporting us into our conclusion: in the universe time has merely the mathematical existence. You continue reading the book if this fact about time is clear to you. If not wait a few days and clear up your mind. It is very important to understand what time is to proceed with our research on the dark side of physics.

The common thinking of cosmology today that the universe has happened in some remote physical time is an illusion. The physical past is non-existent and nothing can happen in something that is non-existent. The universe is running in space only, the universe is time-invariant.

Past and future have only mathematical existence. What has happened 100 years ago has happened in the same space in which you read this book. And what will happen 100 years from now will happen in the same space. Universal space is time-invariant.

No material change in the universe is time-dependent because time is just the duration of changes. Duration, in order to exist, needs to be measured from the side of the observer. Without measurement, there is no duration. In the universe, the time has merely mathematical existence.

The time-invariance of the universe means that imaging universe different as it is today is wrong. The Big Bang is a false model. The universe as we experience it is its real picture. There was no big explosion, there was no inflation period (which is against the first law of thermodynamics), there was no recombination period which was the origin of CMB radiation. CMB radiation is the radiation of universal space which is time-invariant.

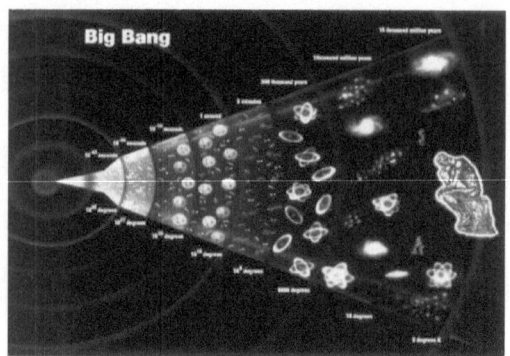

False picture of the universe

The universe is not expanding. Redshift is the result of light losing some energy when pulling out of the strong gravity of distant galaxies.

Let's imagine that the universe is expanding with the light speed from its beginning. In the BB model, the estimated age of the universe is 13.7 billion years which is $4,3 \cdot 10^{17}s$. The radius of the observed mapped universe is 46,6 billion light-years which is $4,4 \cdot 10^{26}m$. To reach the size of today's observed and mapped universe, according to the BB model, the universe should expand since its beginning with the speed of $1,02 \cdot 10^9 ms^{-1}$. The velocity of light is $3 \cdot$

$10^8 ms^{-1} s$. To reach today radius universe should expand with the velocity v which should be 3,34 times bigger than light speed. This shows BB model does not fit into the mapped universe. The velocity of accelerated expansion of the universe today is valued between $6,78 \cdot 10^3 \ ms^{-1}$ to $7,4 \cdot 10^3 ms^{-1}$.

- velocity of expansion accordingly to the BB model $1,02 \cdot 10^9 ms^{-1}$
- velocity of expansion with the light speed $3 \cdot 10^8 ms^{-1}$
- velocity of expansion that is measured $6,78 \ (7,4) \cdot 10^3 ms^{-1}$

The discrepancy between measured velocity of expansion and calculated velocity of expansion accordingly to the BB model (so that BB model could fit in existent measured model) is of the rate $10^6 ms^{-1}$. The BB model seems to be a prediction without experimental data support. BB cosmologists try to defend this discrepancy between the measured diameter of the universe   and   diameter   calculated accordingly to the BB expansion with the proposed Non-Euclidean shape of the universal space (see figure below) which won't work. NASA has measured universal extremely precisely universal space has Euclidean shape.

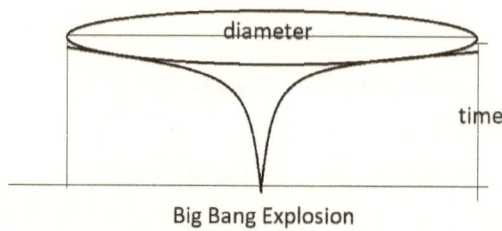

Big Bang Explosion

Inadequate picture of the universe expansion based on Non-Euclidean shape of universal space

Singularities of Big Bang are not bijective
Alan Guth, the founder of the Inflation model said following: "In the inflationary theory the Universe begins incredibly small, perhaps as small as $10^{-24}cm$, a hundred billion times smaller than a proton". Before the size of $10^{-24}cm$ universe was even

smaller. According to Hawking and Hartle, it started from nothing, from a mathematical point. The logical consequence of this scenario is that at the mathematical point energy density and temperature were infinite. With the explosion, the universe started cooling down and expanding. In mathematics, infinity is not problematic. In physics yes, because "infinite temperature + 100 degrees = infinite temperature". Infinity is not a metric term. NASA has measured universe has Euclidean geometry which means universe has infinite volume. We have to understand that infinity of the universal space volume does not mean singularity. Singularities of Big Bang beginning were never measured. Taking the universe as a set X and model of the universe as the set Y we can write the following equation:

$$X: \{Sx\}$$
$$Y: \{Sy\}$$

where between actual universal space denoted as Sx and the model of the space

Sy there is a bijective function. Physical universal space Sx and its model Sy which is Euclidean space are related by the bijective function. Infinite pressure Py in the Big Bang model, infinite density p_y , and infinite temperature Ty have no corresponding element in the set X of the universe. Infinite pressure, energy density and temperature are pure unproved speculation which is not falsifiable. The beginning of the Big Bang is not falsifiable. In Bible universe was created in six days in Big Bang cosmology universe was created in less than second. The elapsed time is the only difference.

In cosmology, only Euclidean space is a proper model of universal space. We cannot use Riemann space or any other spherical geometry because they do not have bijective correspondence with the physical world. Euclidean geometry is time-invariant and also the geometry of universal space is time-invariant. There is no way that today's universal space could appear from a mathematical point as suggested by

Stephen Hawking. There is no need for advanced intelligence to prove the Big Bang model is false. You just put observed data in a computer for systems simulation. The computer will give you "ERROR". Because nothing can expand with the speed of $1,02 \cdot 10^{19} ms^{-1}$. Big Bang model is artificially kept in life because of so much intellectual involvement and money was invested in this theory that nobody is ready to see it is false. This won't help. It is now the time we change this this false and speculative view.

Singularity of space-time inside black hole is not bijective

Maxime Van de Moortel has published an article on arXiv with the title "The breakdown of weak null singularities inside black holes" (2019) where he has developed an idea of space-time having singularity inside the black hole. In the introduction part of this article was shown time has no physical existence which means it cannot be the 4th dimension of space. Only universal space could have singularities. We have to understand that universal space is energy (we call it today "superfluid quantum vacuum" and has variable density. Each physical object is diminishing density of space exactly for the amount of its mass and correspondent energy:

$$\frac{E}{c^2} = m = (p_{max} - p_{min}) \cdot V$$

where E is the energy of the physical object, m is the mass of the object, p_{max} is the density of the space in interstellar space, p_{min} is the density of the space on the centre of physical object and V is the volume of the object. In the centre of black hole density of space is so low that atoms become unstable. Old matter is transforming into fresh energy in the form of elementary particles. Black holes are rejuvenating systems of the universe. There is no singularity of any kind inside the black holes.

2. Big Bang Cosmology and Higgs Mechanism are new Religion of Physics

Science is based on rationality and experiment. Religion is based on irrationality and beliefs. In Big Bang Religion universe has started from an infinitesimally small mathematical point. We know in mathematics that point is dimensionless. Before this point entered into existence according to Big Bang Religion there was nothing. There was only God. There was no matter, no energy, no time, no space. Absolute nothingness.

Why God has decided to created universe Big Bang Religion does not explain. Everything started with the infinitesimally small point which had infinite temperature and pressure. This idea is irrational and non-scientific. Science says that energy cannot be created and cannot be destroyed, it can only be transformed into a different type of energy (first law of thermodynamics).

Beginning of the universe is a religious belief, it is not a scientific hypothesis.

Before the explosion, there was only pure energy. There was supersymmetry (SUSY). All particles were massless. This also is not a scientific theory, this is a religious belief. In science, we know that a given particle in order to exist must have energy which accordingly to the famous Einstein formula E =mc2 is mass. In Einstein vision mass and energy are the same "stuff". SUSY is an irrational religious idea.

In Bijective Physics we know that photon has the energy and corresponded mass, but it has no inertial mass. In the Bijective physics, the mass-energy equivalence principle is extended on the photon. Photon has the energy and so mass, but it has no inertial mass. We can combine two famous formulas of physics, $E = mc^2$ and formula for the energy of the photon, $E = h \cdot v$ where h is Planck constant and v is photon frequency:

$$m \cdot c^2 = h \cdot v$$

And we will get formula for photon energy E expressed as mass m:

$$m = \frac{h \cdot v}{c^2}$$

Photon has energy and so mass, but it has no inertial mass. That's why we called it "massless photon". There is a huge difference between mass and inertial mass which I describe in details in my article "Mass-energy Equivalence Extension onto Superfluid Quantum Vacuum" published 13th of August 2019 in Scientific Reports.

Accordingly to the SUSY in this ideal soup has many massless particles. They have had energy, but they did not have mass. This idea is irrational and religious because in science we know a given physical object which has energy also has mass. SUSY belief is that quarks and leptons have had a superpartner particle as you can see on the picture below.

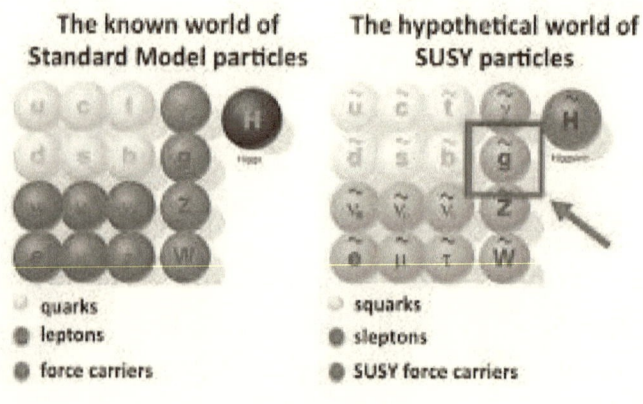

The hypothetical superpartner particles are pure scientific illusion

After the miraculous explosion of Big Bang the superpartners miraculously disappeared, nobody knows where. In every religion, there is a lot of miraculous events nobody understands; the same is the case with supersymmetry and antimatter in Standard model.

The next miracle was the appearance of Highs field. Nobody knows from where it came, but it came and give particles mass. Not all of them, only some particles got mass. This somehow was God's device in the creation of the universe. Higgs mechanism is not falsifiable, it is not bijective, it is against scientific logic, it is the purest form of religious belief. In my article published in Scientific Reports, I explained what is the rest mass, what is inertial mass and what is relativistic mass without Higgs mechanism. I related inertial mass and gravitational mass and describe gravity without graviton. And some grey eminences of physics are complaining about my article.

Giordano Bruno

Except me, nobody is complaining about the fallacy of Big Bang and Higgs mechanism. Because nobody dares to disturb religious feelings related to Big Bang Cosmology and Standard model. These icons are not allowed to be touched. Everybody who will try to show their fallacy will be burned on the fire. If I would live in times of Giordano Bruno, they would burn me on the fire. Now they try to compromise my article. As they have no arguments against, they would be most happy article would be retracted.

Religion to become more convenient is searching for the proofs of their dogmas. Big Bang cosmology is building its credibility on a false interpretation of experimental data.

3. LIGO false interpretation of Gravitational Waves

With gravitational waves discovery we have in physics a wrong understanding that gravity has speed. Why this has happened? Because some people think (also experts) that gravitational waves (GW) are carrying gravity. It is well accepted that GW move with the light speed. But GW are not carrying gravity. GW are the result of matter transformation back into the energy of space (which we call superfluid quantum vacuum), GW are ripples of space which propagate with the light speed. GW have the same speed as light. Because light is a ripple of space and GW also is the ripple of the same space.

In General Relativity (GR) Gravity force is carried by the curvature of space. In Advanced Relativity (AR) gravity is carried by the superfluid vacuum fluctuations which originates in variable density of the vacuum. The formula which bridges curvature of space in GR and variable density of space in AR is following:

$$G_{\mu v} = \kappa \cdot (p_{max} - p_{min})$$

$$in\ units: \frac{1}{m^2} = \frac{m}{kg} \cdot \frac{kg}{m^3}$$

Gravity force is immediate, it does not propagate in space as it is a case with GW. This has to be understood well. Gravity does not work directly between two physical objects, gravity is the result of outer vacuum pressure towards inner vacuum pressure. For example moon and earth diminish density of space and outer vacuum is pushing towards the lower density vacuum. Space is 4D and planets are 3D. They are somehow "locked" into the vacuum. Vacuum has no friction and so planets orbit without the loss of energy.

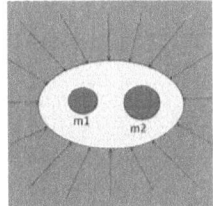

Gravity

Gravity force Fg is equal to centripetal force of the moon:

Fg + (-Fc) = 0.

They work in the opposite direction. GW are changing density of space. Changes density of space is changing space permeability and permittivity. This causes that light changes minimally its speed

$$c = \frac{1}{\sqrt{\mu_0 \varepsilon_0}}$$

Light speed

GW which enters the LIGO interferometer, would be better to say that interferometer is merged in GW is changing a bit permittivity and permeability of space. This changes a minimal light speed and laser beam needs more time or less time to pass the beams. Einstein's idea that GW could shrink or delate space is false.GW is not prolonging or shrinking the LIGO interferometer beams. They are changing permittivity and permeability of space which causes minimal light speed changes. What the measure in LIGE is minimal prolongation or shortage of duration of the light signal motion in the beams. Interpretation of LIGO that variable duration of flight motion in beams is due to GW minimally change the length of the beams of false. How GW which are extremely subtle phenomena could change the iron-concrete basis of the interferometer beam nobody can explain.

4. Why Higgs Mechanism is absolute Failure?

First you need to understand why the idea of Higgs mechanism was born. Photon has no inertial mass and proton has inertial mass. On the base of the fact that some elementary particles have inertial mass was predicted that might be some field pervading entire universe which is interacting with some particles (as for example proton) and slowing them down and is not interacting with some other particles and so not slowing them down (as for example photon).

Higgs field is supposed to interact with some particles, slowing them down and giving them inertial mass. Peter Higgs and other co-creators of Higgs mechanism do not understand the difference between inertial mass and rest mass. Rest mass is expressing the amount of energy which is incorporated in a given physical object.

Inertial mass has physical origin in outer vacuum pressure

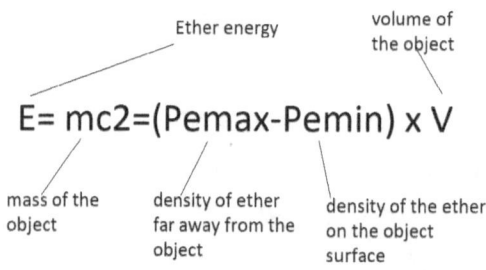

Outer vacuum pressure is the origin of inertial mass

They think that the rest mass is the inertial mass, they do not see the difference which is of the extreme importance to make physics. The result is that the Higgs mechanism is misinterpreting mass as the fundamental element of physics. *Today is widely accepted that Higgs field is giving mass to some particles in the sense of rest mass which means mass as the amount of their energy.* This interpretation is against the mass-energy equivalence principle.

"Mass" and "energy" are the same "stuff", they are related by the formula $E = mc^2$. *Mass is inherent property of a given particle and no field can give mass to the particle*. The entire idea of Higgs field giving rest mass to some particles is absolute failure. Inertial mass of massive particles which is slowing them down has the origin in the diminished density of superfluid quantum vacuum. A given particle with mass **m** is diminishing density of the vacuum in its centre exactly for the amount of its mass **m**.

$$E/c^2 = m = (Pmax - Pmin) \times V$$

where **Pmax** is density of the vacuum in interstellar space, **Pmin** is density of the vacuum on the centre of the proton (or massive object) and **V** is the volume of the proton or massive object. Right side of the formula **(Pmax — Pmin) x V** is the missing part of Einstein's formula which represents mass-energy equivalence principle. I extended his formula on superfluid quantum vacuum (which is the origin of universal space) and so the enlarged formula is showing the origin of inertial mass. Inertial mass is the physical property of the rest mass of a given physical object which is the result of outer vacuum pressure. As Peter Higgs and Higgs mechanism co-creators think that inertial mass is rest mass we have now the situation nobody knows what is mass. They do not understand*: **Rest mass is expressing the amount of the energy incorporated in given physical object at rest. Outer vacuum pressure on rest mass is the physical origin of inertial mass of a given physical object.** With Higgs mechanism **mass** has become something very abstract and speculative. The result is, they are some false ideas of **negative mass** existence in today physics.

Albert Einstein and Max Planck are turning in their graves. 95% of articles in theoretical physics today are failure, and my article is meant to be removed because some unknown "experts" who have no courage to tell publicly their names do not understand physics.

In my article published recently in Scientific Reports I have shown the difference between inertial mass and mass as the amount of energy. I have shown that introduction of Higgs mechanism is not necessary, it is not adding anything to physics, it is complicating physics to the extend that today nobody knows what is the mass of an elementary particle.

Seems that cyclotron lobby is trying my article to be removed. They are not happy with the idea that God particle is dead. The entire cyclotron physics is the absolute failure. I know it is painful, but to develop physics, we have to admit mistakes. Big Bang cosmology, Supersymmetry, inflation model, are dead branches of physics.

Albert Einstein has proved inertial mass and gravitational mass of a given physical object have the same value. I proved they have the same origin, namely, in the variable density of superfluid quantum vacuum which before 1905 was called "ether".

5. Ether Model is bridging Special Relativity and General Relativity

Special Relativity was born in 1905. General Relativity was born in 1915. For 10 years Einstein has studied how to incorporate gravity in the Special Relativity Model. He came to the genius idea: on the Earth centre we have gravity force and gravity acceleration 9,8 ms2. If someone would go in space where there is no gravity what would happen? The gravity acceleration in his spaceship at rest would be zero. But when put on the engine and accelerating with the 9,8 ms2 the astronaut would have the same experience as on the Earth centre. This thought experiment is elegant, it makes sense and it was excepted by scientific community.

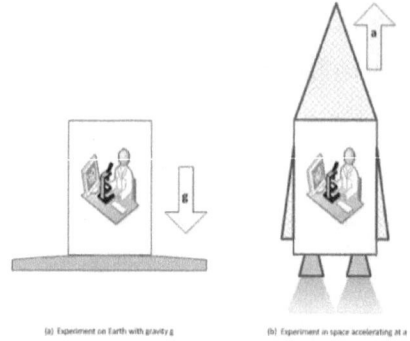

(a) Experiment on Earth with gravity g   (b) Experiment in space accelerating at a

Thought experiment of Albert Einstein

This thought experiment has big implications, namely, it shows that inertial mass and gravitational mass are equal. In this way Special Relativity was fully incorporated in General Relativity.

But the open question still was: *What is the origin of inertial mass and of gravitational mass?*

Considering that universal space is empty there was no possibility to find out why inertial mass and gravitational mass are

equal and also no what is their origin. With the ether model this was recently done in my article published in Scientific Reports. In this article I used more convenient name for ether: *superfluid quantum vacuum*.

Every physical object is diminishing density of the ether in its centre exactly for the size of its mass. Because of its variable density ether is exerting push towards the centre of the object. As physical objects are 3D and ether is 4D (see previous stories) somehow physical object is pushed equality from all directions. This action of ether on physical object is the origin of inertial mass and also origin of gravitational mass.

Variable density of the ether is physical origin of gravity potential. With simple words: gravity is in the space, gravity is the very structure of space (where "structure" means variable density). Let's say 10 meters above Earth centre: is nothing is there, nothing will be attracted. But when a physical object is there, it will be pushed toward the centre.

In physics today is not clear yet that inertial mass is not rest mass. Inertial mass is the result of ether push towards the given physical object. And the amount of the energy of ether which is incorporated in this object is its rest mass.

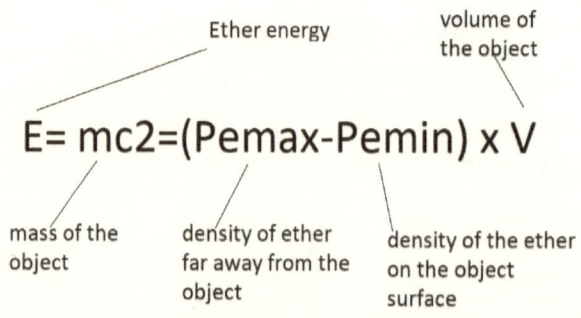

In the formula above **Pemax** means energy density of the ether far away from the object and **Pemin** means energy density in the object centre.

We can use both terms: **density of vacuum (P)** and **energy density of vacuum (Pe)**

Pe = P x c2

Both expressions are right. Energy density is sometime more convenient, because is shows the missing right part of Einstein **E=mc2** equation.

When a given physical object stats moving in ether it will interact with ether and absorb some of its energy. This energy is the object kinetic energy (see previous stories on ether).

Introducing ether back into physics we do not needs Higgs mechanism, all works perfectly. Introduction of Higgs mechanism in physics ids the result of ether abolishment and is not necessary. It seems that entire cyclotron physics has no future. Bringing ether back into physics everything can be described without exotic particles which have lifetime from E10−10 seconds to

E10–25 seconds. My opinion is that this "particles" are artificial manmade and do not exist in physical universe in their own.

Today academic science is often lost in theoretical speculations which have no contact with the real world. This is the weak point of today science. Every exotic thinking which has some support in mathematics is thought to be science. In this way science is becoming pseudoscience.

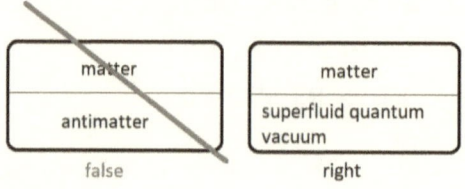

The idea of antimatter existence is false

Albert Einstein told us: **I do not believe in mathematics**". We should take his quote seriously. Math is a good tool, but it can not guide physics. Physics is the queen and math is the servant.

Albert Einstein

In the universe ether is the complementary element of matter. In today physics we call ether with the new name **superfluid quantum vacuum**. Ether which is the physical origin of universal space has no entropy, matter has entropy. In the universe 95% of the energy in from of ether is syntropy type and 5% of the energy in the form of matter is entropy type of energy. Antimatter is false concept, it is non-existent in the physics universe. Antimatter research will never give any technological application. Big Bang model and antimatter model are dead branches of physics.

6. **CERN Antimatter Research is a scientific Illusion**

In 1928, British physicist Paul Dirac wrote down an equation that combined quantum theory and special relativity to describe the behaviour of an electron moving at a relativistic speed. The equation — which won Dirac the Nobel Prize in 1933 — posed a problem: just as the equation $x2 = 4$ can have two possible solutions ($x = 2$ or $x = -2$), so Dirac's equation could have two solutions, one for an electron with positive energy, and one for an electron with negative energy. But classical physics (and common sense) dictated that the energy of a particle must always be a positive number.

Dirac interpreted the equation to mean that for every particle there exists a corresponding antiparticle, exactly matching the particle but with opposite charge. For example, for the electron there should be an "antielectron", or "positron",

39

identical in every way but with a positive electric charge. **The insight opened the possibility of entire galaxies and universes made of antimatter.**

But when matter and antimatter come into contact, they annihilate — disappearing in a flash of energy. The Big Bang should have created equal amounts of matter and antimatter. So why is there far more matter than antimatter in the universe?

At CERN, physicists make antimatter to study in experiments. The starting point is the Antiproton Decelerator, which slows down antiprotons so that physicists can investigate their properties **(this text above is on CERN home page).** This is the official story of physics today. I think that the idea of antimatter is highly exaggerated and false. Above we can read following: **The insight opened the possibility of entire galaxies and universes made of antimatter.**

Lets start with the positron which is the antiparticle of electron. It has been found in cosmic rays, but in the contact with ordinary matter is unstable. It has lifetime about 10E-10 seconds. Do positron deserve to be called "particle" in the sense that it could be a consistent element of illusory antimatter? In my opinion positron is not more than the momentary flux of energy released in cosmic rays. It immediately disappears back into the energy of the vacuum. This is what all experiments which have been done in past 80 years are confirming. **Positron is a momentary flux of energy**. *The idea, that the position could be a consistent part of some exotic antimatter is for now only unproven speculation.*

Richard Feynman has proposed that electron going back in time is positron: The **one-electron universe** postulate, proposed by John Wheeler in a telephone call to Richard Feynman in the spring of 1940, is the hypothesis that all electrons and positrons are actually manifestations of a single entity moving backwards and forwards in time. According to Feynman:

" I received a telephone call one day at the graduate college at Princeton from Professor Wheeler, in which he said, "Feynman, I know why all electrons have the same charge and the same mass" "Why?" "Because, they are all the same electron!"

Back in 1940, the idea of time as the 4th dimension of space was the mainstream idea and so the bizarre idea of Feynman was taken seriously. Today we know that time has merely the mathematical existence.

Between Feynman proposal of what positron could be and observation of the positron in cosmic rays there is a huge discrepancy. This would mean that cosmic rays are moving back in time. This is very good example how famous physicists can have very bed ideas.

Antiprotons also are found in cosmic rays. The experiments confirm antiproton has life time about 32 hours. Positron has no stable lifetime, antiproton has no stable lifetime. How they could be a composite elements of something we call **antimatter?** Ordinary matter is build our of protons and electrons which have unlimited life time. That's why matter is stable and existent.

Only stable elements can build a stable structure. This fact in CERN is not understood yet. They are searching for antihydrogen. Antihydrogen which should be should be composed out of positron and antiproton is pure illusion.

Yes, might be they will find antihydrogen. It will have a lifetime like quarks which is between 10E-23 to 10E-25 second. And again the Nobel will be delivered for something that does not exists. Yes, entire cyclotron physics is failure. How long the mainstream physics will need to see this illusory research on antimatter makes no sense nobody knows.

In the universe there is a primordial symmetry between matter and superfluid quantum vacuum. Accordingly to this symmetry every physical object with mass is diminishing density of superfluid quantum vacuum accordingly to its mass. We have two fundamental elements in the universe: matter/energy and quantum vacuum. Accordingly to the bijective research methodology we can see universe a set a **U** with two elements: matter/energy element which value is 1 and superfluid quantum vacuum which value is zero (0).

U : {0, 1}

in set theory zero is empty set: **0 = { }**

and 1 is set with the number zero: **1 = {0}**

Out of this follows: **U : {{ }, {0}}**

Matter/energy element represented as 1 is manifested element of 0 which is non-manifested superfluid quantum vacuum.

Universe is a system which functions on the basis of the set theory. That's why the development of computers which work on 0 and 1 was possible. Computer technology is the expression of the intrinsic property of the universe complementary duality: 0 and 1. Antimatter research will never have technological implementations because it is an illusion. Antimatter does not exist in the physical universe. The complementary element of matter/energy is the superfluid quantum vacuum.

Electromagnetism is the excitation of superfluid quantum vacuum which before 1905 was named **ether.** The result of the research how matter and electromagnetism are interrelated is radio, TV, calculators, computers, your PC. The next step of research should go into direction of searching the relation between matter and variable density of the vacuum. The result will be antigravity. If we manage to polarize vacuum we will get free energy. I think Tesla had developed this technology, but having energy for free was not okay for Morgan. So the Tesla tower was destroyed.

Tesla tower was producing free energy from vacuum

Antimatter research and exotic particle research in cyclotrons will never have any technological implications. It is the biggest budget in the history of mankind spent on the research in which theoretical predictions are false and searching for them is a waste of time and money.

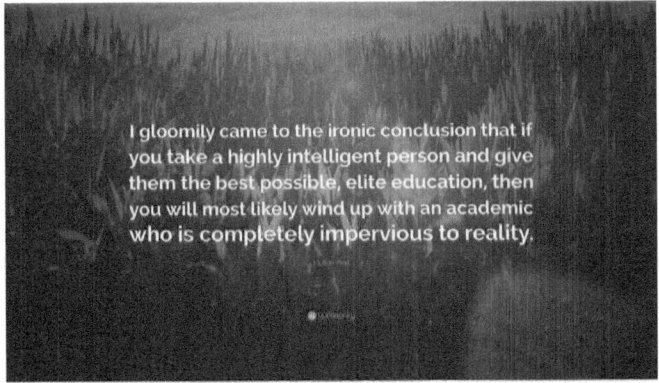

I gloomily came to the ironic conclusion that if you take a highly intelligent person and give them the best possible, elite education, then you will most likely wind up with an academic who is completely impervious to reality.

Halton Arp, American physicist

To prevent this disaster of today physics being completely impervious to reality I develop Bijective Research Methodology in Physic.

7. Ether and fallacy of Nuclear Physics

Unification of strong nuclear force and gravity

In bijective physics a given element can be a constitutive element of some more structured physical object if it is stable, it means it has stable lifetime. We have 2 stable particles in the physical universe: **proton and electron**. One should ask: Why proton and electron are stable? The answer is: they are both vortex of the **ether** which we call today in **physics superfluid quantum vacuum.** That proton and electron can be seen as vortexes of ether has scientific basis in Dr. Sbitnev article here: https://arxiv.org/abs/1603.03069

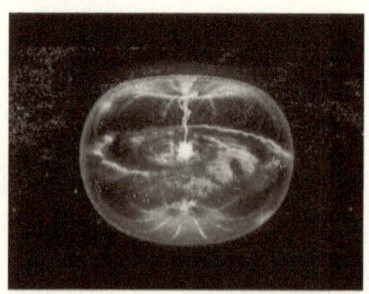

Ether vortex

Ether is non-created syntropy type of energy, the physical background of universal space. As ether is syntropy type of energy proton and electron practically have infinite lifetime.

Neutron is not "particle", neutron is not stable. When isolated will decay into proton and electron in about 15 minutes. In atom nucleus we have only protons and electrons.

	orbit	nucleous
H	1 electron	1 proton
He	2 electrons	4 protons + 2 electrons
Li	3 electrons	6 protons + 3 electrons
Be	4 electrons	8 protons + 4 electrons
B	5 electrons	10 protons + 5 electrons
	N electrons	$2N$ protons + N electrons

Electrons and protons in an atom

Elements with high numbers of protons in the nucleus become unstable as for example uranium. This instability of elements with high number of protons is the result of complexity. We call it in bijective physics **complexity instability**. This is natural law that some structure is stable in the limits of complexity. Once something is too complex it will become instable and fall apart in less complex system.

Standard model says that W boson and Z boson are caring weak nuclear force. This is not true. The weak nuclear force does not exist at all. The instability of atoms with big atomic numbers is due to complexity instability of the atoms which have big atomic numbers. And the famous "God particle" is not giving mass to the elementary particles. Higgs boson is artificially made in cyclotrons and has no existence in physical universe.

Inside the nucleus protons and neutrons (as composite particles) are attracted because they diminish density of the ether which creates gravity force. **Strong nuclear force is gravity force inside the nucleus.** The article with calculations of diminished density of the ether on the proton centre, Earth sentre and black hole centte I published 13th of August in Scientific Reports with the title: "Energy-mass Equivalence Principle Extension on Superfluid Quantum Vacuum". Density of the vacuum in the proton centte is smaller than on the Earth sentre. This is why strong nuclear force has such a power in comparison with gravity force.

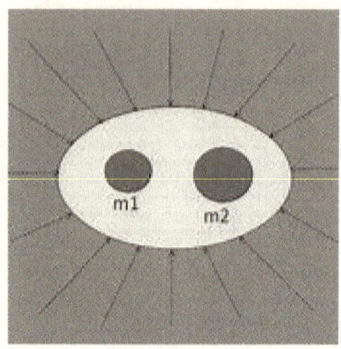

Ether pressure carries gravity

Both forces work on the same principle. Two physical objects are diminishing density of the ether. The outer higher pressure of ether is pushing towards the lover pressure in the area of two physical objects. This is valid from microscale to the macroscale, from proton to stars.

Existence of quarks inside the proton is pure fantasy. All the particles "discovered" in cyclotrons are just momentary fluxes of energy released by proton collisions. They immediately integrate back into ether.

Top quark	171,2 GeV/c2	10E-23 s	1995
Higgs boson	125 GeV/c2	1,56 x 10E-22 s	2013
Z boson	91 GeV/c2	3 x 10E-25 s	1983
W boson	80 GeV/c2	3 x 10E-25 s	1983
Bottom quark	4,2 GeV/c2	10E-23 s	1977
Charm quark	1,27 GeV/c2	10E-23 s	1974
proton	938 MeV/c2	stable lifetime	1886
strange quark	104 MeV/c2	10E-23 s	1968
Down quark	4,8 MeV/c2	10E-23s	1968
Up quark	2,4 MeV/c2	10E-23 s	1868

Artificial particles which have no existence in physical universe

All these particles have life time between 10E-23 s and 10E-25 s. How this "particles" could build something stable if themselves are unstable Standard model has no answer. This is a very deep shock for physics. I'm sure they will be a huge resistance to this story, but facts are facts. All cyclotron job is pointless. It will never have a technology application because all is made up only in the human mind. All these particles are artificial and do not exist in physical universe.

There was no some magic time of supersymmetry, there was no inflation of the universe and no Big Bang. Sonner we acknowledge this better for physics progress. To make mistakes is inevitable part of human life. It is also valid in science. It is normal. What is not OK. is holding on the old ideas despite clear evidence that they are wrong. This is the weak point of today physics. I hope things will progress well. **It is time to switch into the new paradigm.**

There is no antimatter in the physical universe. Antiproton is artificial manmade flux of energy and has no existence in the physical universe. In the universe we only have matter and ether. All the rest (supersymmetry particles, antiparticles) exist only in the human mind. We have only two particles in the universe which are proton and electron. Both are vortexes of ether.

System Stability Model and Fallacy of Nuclear Physics

Life is the best prove of what is real and what is an illusion. If one tire of your car is broken, you can not drive. The system has 4 elements (tires) and when one is not stable, the entire system becomes unstable. You work in a team where one person is psychologically unstable, the solid troublemaker. The entire team will become unstable and will not function well. You build a house and you have 4 basic pillars. If one pillar is not well done the entire system (house) will become unstable and will collapse. The stability of solar system is based on the stability of all planets. Imagine that one planet would explode. The entire solar system would become unstable.

$$S = \sum_{i=1}^{N} E_i$$

A given system S is stable when all elements E in the system are stable

Life is teaching us: **A given system S has a potential to be stable only when all its elements E (E1, E2, ….En) that build a system are stable.**

Neutron is built out of proton and electron which are stable, and itself is not stable. In Standard model we have a fancy idea that unstable elements can build a stable system. Quarks are highly unstable, their life time is extremely short and they are supposed to build proton which is stable, it has unlimited life time.

Nobody has an explanation of how this is possible. Nuclear physics has established some fancy believes about how unstable quarks can be constitutive elements of the proton. Nobody has an explanation of how this could work and seems nobody cares. Nuclear physics lobby is so strong that it does not need to care about the common sense of Standard model. They are living in their imaginary world, planning bigger cyclotrons to discover not only "God particle" but also "Angel particle", "Holly Ghost particle", and all kind of other exotic particles. And Nobel prizes will be delivered and big fame and success will follow. All this is the hypocrisy of today nuclear physics. For now, nobody is ready to see this, I'm the early bird singing the song of truth.

You remember the child saying in *Hans Christian Andersen* story: **The king is naked!**

This is where we are in physics today. Big Bang is naked, supersymmetry is naked,

God particle is naked, Cyclotron physics is naked!

8. Re-examination of Physics with Bijective Research Methodology

Physics is my daily occupation for last 35 years. We have internet and I read hundreds of articles on arXiv which have been published in important journals of physics. My general observation was that that complexity of physics is increasing with time. Each year I read articles which have been more complicated and more difficult to understand. ***Despite my knowledge of physics was increasing I understood less.***

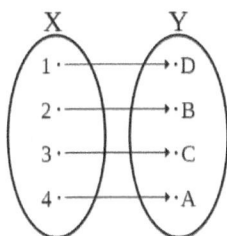

Bijective function of set theory

I could not grasp why, but somehow my intuition told me that too much complexity makes every system unstable. We can observe this by atoms with high atomic numbers. Atoms nucleus with number 90 and higher are unstable. They fall apart into atoms which have less protons in the nucleus.

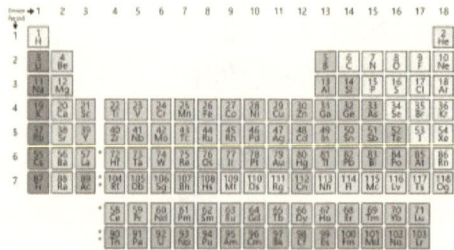

Table of Elements

I start thinking maybe physics is entering the similar state of **complexity instability.**

To make this natural process more efficient I decided to develop a methodology which will eliminate in physics all that might be false. In this way physics will diminish complexity and increase clarity, it will become more **adequate model** (or we can say *picture)* of physical reality. In this my endeavour set theory was of the best help: universe was taken as a set **X** and model of the universe was taken as a set **Y**. Set X and set Y are related by the bijective function which means that every element in the set X has exactly one element in the set Y.

A bijective function,

$$f: X \rightarrow Y,$$

where set *X is {1, 2, 3, 4}* and set *Y is {A, B, C, D}*.

For example, $f(1) = D$.

By using bijective research methodology we should decrease physics complexity and increase its clarity. This was my dream since my early age when I read the book of Stephen Hawking "The Brief History of Time" where he explained appearance of energy in the universe with mathematical trick saying that energy of matter is positive and energy of gravity is negative and that both energies are multiplying in the first moments of Big Bang. I was attending 5th class of primary school that time and I decided when I will be adult, I will fix this inappropriate ways of making physics. You cannot explain appearance of energy in this way.

Bijective research methodology excludes existence of negative gravity. Hawking idea which also is Alan Guth's idea we can mathematically write as:

$$nEm + (-nEg) = 0$$

At the very beginning, n is zero and is increasing in inflation. Both of the energies are increasing and their origin still is not explained. Saying that their sum is zero is not solving anything.

Energy of gravity having negative mathematical sign does not pass test of bijective research methodology.

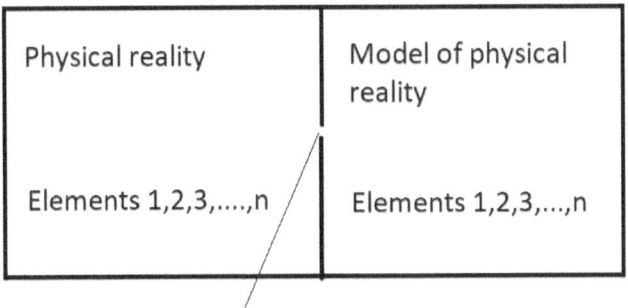

Physical reality	Model of physical reality
Elements 1,2,3,....,n	Elements 1,2,3,...,n

test of bijective research methodology

How to stat with re-examination of physics? I decided that I will use only elements in the model which I can directly observe with my eyes or directly observe their manifestations.

The first element is matter, we all observe matter, the second is energy (electromagnetic energy) which also we all observe, the third is change and the forth is space in which change exist, the fifth is me as the observer. Time I did not take as an element because time we cannot observe with senses. We can only observe change. In this perspective time is the mathematical numerical order of events running in space. This already was Einstein vision on time.

We have four fundamental elements in the universe which we can observe (in generally we can say: which the observer can observe) with senses (sight):

1. matter (M)

2. energy (electromagnetic) (E)

3. change (C)

4. space (S)

The fifth element is the observer.

5. the observer (O)

Matter, energy and change exist in space. In this perspective we can define space as the medium in which changes of matter and energy occur. We called this medium in physics ETHER which was abolished back in 1905. As space contains energy and matter we decide that space also must be a kind of energy. We are back to the ETHER. We have now in the set X four element and we have for elements in the set Y:

X : {Sx, Ex, Mx, Cx, Ox}

Y : {Sy, Ey, My, Cy, Oy}

With these 5elements we can describe entire universe. Energy of ether that today we call "superfluid quantum vacuum" is the energy of dark matter and dark energy. If ether would not be thrown out of physics there would be no need to introduce dark energy and dark mater into physics.

9. Misunderstanding of Physics Objectivity

Physics view on reality is not objective, it is rational which is far from being objective. The common view we have about science is that what is "scientific" is real, is objective. This is a big misunderstanding. What we are doing in physics is that we are building model of reality which are pictures of the world. The picture is never the world itself. Pictures we make in physics are rational pictures, designed by the rational scientific mind.

Immanuel Kant

The common belief is that physics is telling us the truth about reality which is pure illusion. Reality is what Immanuel Kant called "Ding an Sich" which means the word as it is without being interpreted by your mind. Physics and science in generally is interpreting the world. The interpretation is rational which is far from being objective.

OBJECTIVE = DING AN SICH = WORLD AS IT IS (without mind interpretation)

In generally, human experience is indirect, between perception and experience there is a mind elaboration.

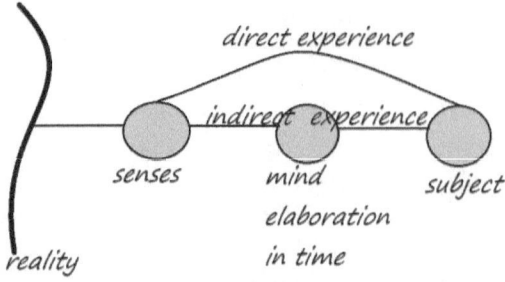

1. Information of reality comes into senses.

2. Mind elaborates information

3. The subject experience information.

In physics the elaboration of information is rational and so the experience is rational and far from being objective. What means **objective**? It means not influenced by the mind. The subject which is fully aware the way his mind elaborates information has possibility to experience reality without mind influence.

Meditation

This is possible only by practising meditation. Meditation is taking you out of the mind and is placing you into the subject. Once you are the subject, you see you are not the mind. You can experience reality as it is coming into your senses. You have direct experience of reality which is essential for the survival of human species.

Only meditation can take us our of the illusory world we are living in. *Scientific experience of reality is a kind of rational illusion*.

Science is about the quantity of the world, meditation is about the quality of the world. Without knowing the quality we cannot survive, we will destroy the life on this planet in the name of the profit. Only meditation will wake up human being from its limited mental picture of the world and give him broader experience of the world where all is interconnected.

Once you know that you are the part of a huge universal process which we call **LIFE**, you are automatically **ecologically awaken**. You will not poison the soil anymore, you will not buy things you do not need, you will take care of nature knowing that nature is the basis of your life.

10. "Out of the Box" thinking" to Progress Theoretical Physics and to Discover the Purpose of your Life

It is modern for the last 20 years to use this term. It means to think in a new original way and get good solutions. The main problem is that the idea has arisen to make more money. **Out of the Box Thinking to make more money globally won't work. Out of the Box Thinking to build a sustainable society will work.** Our idea of profit is the main barrier to the progress of human society. There is no profit in the universe, there is no profit in nature, profit exists only in the human mind. **Out of the Box** means you start thinking about what you can do that this world will be more beautiful, more happy, more healthy.

Out of the Box Thinking

You can do every day a beautiful practice, very useful for out of the box thinking. You sit silently, close your exes and imagine you are far away from the Earth in the vastness of the universe. You see our planet floating in the universal space.

You ask yourself what are your best skills and qualities to add something to the beauty of life on this planet. 100% you will receive strong inspirations from the cosmic intelligence. If you do not believe in cosmic intelligence if you think the only purpose of life is making money and spending it than you will remain in the box. If you are happy like this, it is perfectly OK. If you are not happy than the practice will change your life. **Real is what works.** Your intention to make this world more beautiful works.

If you are a scientist, especially a theoretical physicist, you start watching your mind. Find a near Buddhist centre and do a 3-month vipassana course. By watching your breathing you will increase abilities to watch your mind. You will step out of your **"mind-box"** and you will see that your psychological time is the bigger box in which is your mind-box. After a while, you will be able to take your **mind-box out of your time-box** and d a fresh vision will open. You will see that in your models time is just a mathematical operator of motion of elementary particles, massive bodies, and stellar objects. You will enter what Julian Barbour calls "The Third Revolution of Physics".

English physicist Dr. Julian Barbour

Once you are out of the two boxes (the first box is **mind-box**, second is **time-box**) you will find the most exciting adventure of your life in physics: **Bijective Physics.**

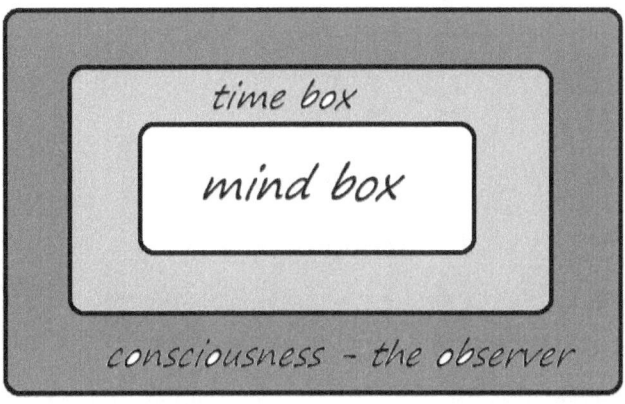

You will become the **Conscious Observer**, you will merge into the secrets of the universe far beyond your today most wild thought and crazy theory. You will discover that in the universe all is perfect, they are no problems. All the problems in today physics are showing our model are not right. You will drop re-normalization methodology because they will be no need of it. You will step out of the religion of Big Bang and Standard model.

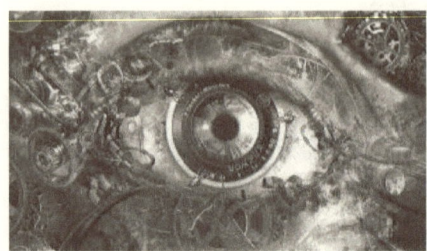

The observer

The observer is the core of physics. He/she is observing the world via senses, he is using his mind building the scientific model of physical reality, he is designing the experiment to prove or disprove the model.

We have to understand that the observer is not the mind. Mind is thinking the observer is aware of the thinking process. We have two ways of thinking:

- thinking without observer's supervision (1)

- thinking with the observer's supervision (2).

The difference between (1) and (2) is huge.

In (1) the observer is unconscious, he is merged in the thinking process. He is developed to the point: "I think, So I'm" (Cogito ergo sum).

In (2) the observer is conscious of the ways his mind works. He is developed to the point: "I'm aware of my mind activities, so I'm".

conscious observer	unconscious observer
aware of the way mind is thinking	unaware the way mind is thinking
not identified with the mind	identified with the mind
free of the mind	imprisioned in the mind

Conscious and unconscious observer

The conscious observer (2) has higher cognitive ability of being aware the way mind works. The unconscious observer (1) is not aware what his mind is doing. He is identified with the mind in good and in bad.

Conscious observer is totally free of the mind. He/she uses mind as a tool. The unconscious observer is the slave of his mind. The mind is the boss, the observer is the servant.

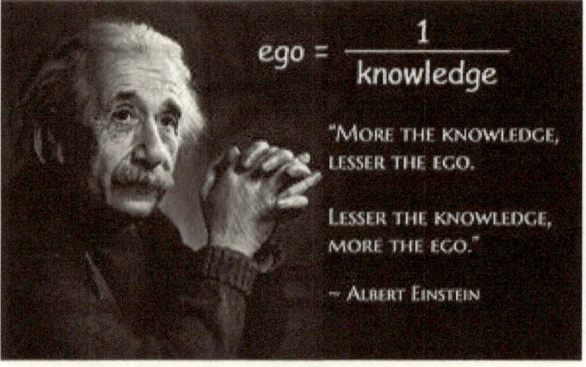

The conscious observer is free of what we call in psychology "ego", the unconscious observer is the slave of the ego. Conscious observer is not identified with the data his mind purchases (mind's knowledge). Unconscious observer is fully identified with the data (mind's knowledge) his mind purchases. Conscious observer has cosmic knowledge, unconscious observer has only human knowledge. Conscious observer is entangled with the entire universe, unconscious observer is entangled only with this mind. He lives and experiences universe in his small "MIND-BOX".

Conscious observer is building physics on experimental data, unconscious observer is interpreting the data in the way that his ideas are confirmed. Big Bang cosmology and Standard model are school examples how unconscious observer is destroying the beauty of physics. 2020 is the end of Big Bang Cosmology and the end of the Standard model. I work on this every day, because "enough is enough". We have to save physics from becoming silly. It is not right that thousands of students are listening the lectures on the Big Bang cosmology and Standard Model which are both an utter misunderstanding.

Only the conscious observer can progress physics. The conscious observer is intelligent. The unconscious observer is only pretending to be intelligent. The conscious observer is not emotionally bound to his knowledge. The unconscious observer fully identifies with his knowledge. He/ she is like a child, he will not give his toy away. Big Bang cosmology and Standard model are the toys built by the unconscious observer (see my other stories on these subjects).

Albert Einstein

The unconscious observer experiences the physical reality in the frame of his psychological time. The conscious observer is free of psychological time. Albert Einstein was free of psychological time. Max Planck was free of psychological time, Ervin Schrödinger was free of psychological time, Julian Barbour is free of phycological time, and I'm free of psychological time. The entire universe is time-invariant.

Conscious observer is the only way to progress physics. Bijective Research Methodology which is build by the conscious observer. Bijective Research Methodology confirms Big Bang Cosmology and Standard model are groping in the darkness of the human mind. They are not falsifiable, they are based on wrong interpretation of data, they are pure failure. For how long this will go on nobody knows. Human intelligence is like a Schrödinger cat. You never know, cat is alive or dead.

The conscious observer is building physics free of the illusions of the unconscious observer. You start observing (watching your mind) 15 minutes a day. You will see in a few months time you will be able to see how your minds works. Your mind will become your servant. He will do the job you ask him. He will not invent irrational ideas as for example "negative gravity", "chronological time protection", "negative mass", and so on. This is where true physics starts.

11.   Bijective Physics solves Olber's Paradox

In astrophysics and physical cosmology, **Olbers' paradox**, named after the German astronomer Heinrich Wilhelm Olbers (1758–1840), also known as the "**dark night sky paradox**", is the argument that the darkness of the night sky conflicts with the assumption of an infinite and eternal static universe. In the hypothetical case that the universe is static, homogeneous at a large scale, and populated by an infinite number of stars, then any line of sight from Earth must end at the (very bright) centre of a star and hence the night sky should be completely illuminated and very bright. This contradicts the observed darkness and non-uniformity of the night.

Olbers' paradox

Why is the night sky dark?

If the universe is infinite and ageless (according to Newton's model), the night sky should be uniformly bright since there will be stars in every direction. If stars are further, away they will be less bright but more numerous.

The brightness decreases according to the inverse square law, but the number of stars is proportional to distance cubed.

SOLUTIONS

The universe is **not** infinite in size.

The universe is **not** infinitely old. Some light has not yet arrived.

The universe is **expanding** so some of the light has been stretched (red-shifted) outside the visible spectrum.

The proposes for Olber's paradox solutions

Many solutions to solve the Olber's paradox are proposed, but all are false. You can read them above. We will find now the solution for Olber's paradox which will be based only on astronomical observations.

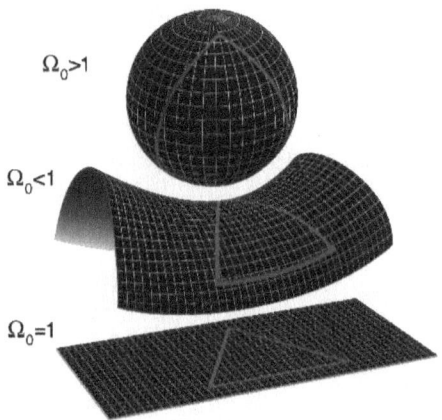

$\Omega_0 > 1$

$\Omega_0 < 1$

$\Omega_0 = 1$

Universal space has Euclidean shape

NASA has measured back in 2014 universe space has Euclidean shape. They measured angles between three stellar objects and they have seen the sum of the angles forming the triangle is always 180 degrees. This means Universe has Euclidean shape and is infinite.

To solve Olber's paradox we have to understand first why we have days and we have a night. This is because Earth is rotating around its axis.

night is the shade of the Earth

Earth

Sun

Night is when we are in the shade of the Earth

Whey you are on the Sun side, you have day. When you are on the other side you have night. Night means you are in the shade of the Earth. Now you imagine that Earth stop rotating when you have day. You will have only day, night will be gone.

You decide to find another planet and you leave in a fast spaceship. You will need a few month to come far away from the Sun to have night again. Going away from the sun its luminosity will decrease. First days the light will become weaker, after a whole you will be in darkness. Entire universal space is dark. The formula for the brightness of the star is following:

$$b = \frac{L}{4\pi d^2} \qquad\qquad L = (4\pi d^2)\, b$$

- the distance **d** to the star,

- the apparent brightness **b** of the star, and

- the luminosity **L** of the star.

With the rising distance from the star the brightness is diminishing. First days on your travel away from the Sun you have been always in day, than slowly light was less, finally you enter darkness.

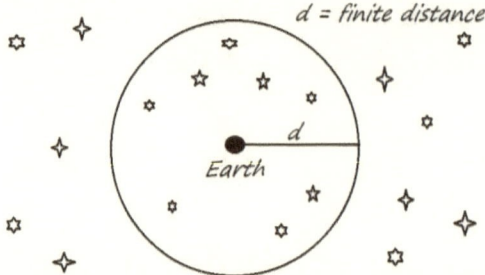

Brightness of the stars which are on finite distance

Why we have night when being on the other side of the Sun? Because brightness of the stars which are in the area of finite distance from us is too low to make us a day when we have a night.

This is solution of Olber's paradox based on astronomical observations of universal space being infinite. The stars which are on the infinite distance have no impact on us. Their light never reach Earth.

12.   Black Holes are rejuvenating Systems of the Universe

The old idea was that black holes are sucking in the old matter and matter is miraculously disappearing behind the event horizon. This view on black holes is a bit mysterious and against the first law of thermodynamic: energy cannot be created and cannot be destroyed. The old idea also was that in the centre if the black hole there is a singularity, infinite pressure and temperature.

The new understanding of what is happening inside of the event horizon of black holes is more realistic. In August 2019 I published an article in Scientific Reports here I also calculated density of universal space (we say in physics "superfluid quantum vacuum") in the centre of the proton, in the centre of our planet Earth, in the centre of the neutron star and on the centre of the black hole. Surprisingly, density on the proton centre is smaller than on the Earth centre but still much bigger than on the centre of the black hole. This excludes existence of mini-black holes which today is often discussed in popular physics journals.

After getting these calculations i came on idea, that on the centre of the black hole atoms become unstable because of low value of space density. Imagine yourself on the iron net 10 x 10 com.

Standing on this net you will be stable, but when distances increases on 20 x 20 cm you become instable. When distances are 100 x 100 cm you become unstable, you fall through the net and you fall appart.

Something similar happens on the Black hole centre, atoms are loosing the stability because they are in the very low density space. They fall apart in elementary particles.

Applying "Newton shell theorem" in calculations of space density inside the black hole, we see that just after event horizon density of space is increasing and than slowly decreasing going towards the centre of the black hole.

In the centre of the black hole we have the same circumstances as on the centre. Matter is falling apart into fresh energy in the form in of elementary particles and cosmic rays. When this pressure inside is becoming stronger as gravity force, the black star is exploding in supernova.

In the centre of galaxies we have supermassive black holes which have a hole in the centre. In the process of their creation the pressure in the centre was creating the hole along the rotational axis. These black holes have a jet of elementary particles being continuously expulsed in the universal space because of high pressure inside the hole along axis.

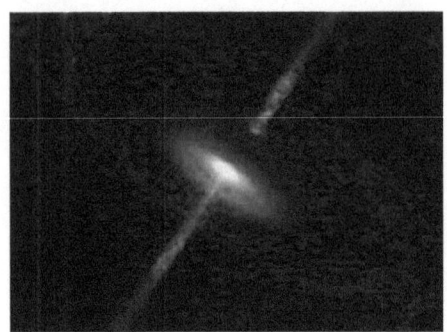

Jet of elementary particles

In the centre the matter is continuously falling apart in elementary particles and this process created a high pressure which is creating the yet of elementary particles. In this sense black holes are transforming old energy in the form of matter into fresh energy in the form of elementary particles. Black holes are keeping entropy of the universe stable. They are rejuvenating machines of the universe.

This is the new model of how universe works. The days Big Bang are running out. In the universe time has only the mathematical existence. This was already said by Einstein but he was not well understood.

Understanding that time has only mathematical existence it is clear that cosmological principle is time-invariant. This means that universe on the large scale has eternally the same face. Vastness of space filled with innumerable galaxies which are continuously recreating universe. Sounds much better, more elegant and with much more sense than magic mysterious explosion of the Big Bang which requires existence of God.

There is no God beyond the universe. I know, physicists will not be happy with this sentence, religious people will not be happy. But this is the truth.

13.   Albert Einstein on Time and NOW

Regarding time, Albert Einstein says three very essential things:

1. **Past and future are stubbornly persistent illusions.**

2. **There is no other time that order of events by which we measure it.**

3. **.....there is something essential about NOW which is out of the realm of science.**

Einstein was aware that ETERNITY IS NOW. He could not said this and so here prepared everything for the next generations to find out themselves.

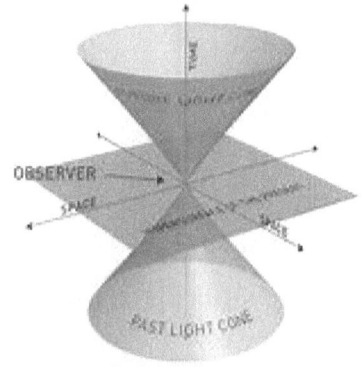

Minkowski space-time model

In Einstein's Special Relativity Theory hypercentre is the present, the NOW. Time as 4x = ict is imaginary coordinate. i on square is -1, so the 4th coordinate is not real. This NOW, is eternity itself. Changes run in space which is timeless. We experience timelessness of space as NOW. Past and future are only in the mind. Julian Barbour is fully aware time has no physical existence. Juliann says that the right understanding of time represents **The third Revolution of Physics.** He is 100% right.

Noether's theorem states that every differentiable symmetry of the action of a physical system has a corresponding conservation law. A given physical system is symmetric in time if its entropy remains unchanged.

What does it mean **Symmetric in time**? Here we give the advanced understanding of Noether's theorem, namely: **a given physical system is symmetric only in space (never in time) if during its duration its entropy remains the same**. In physics, we experience the increase of entropy and run of physical changes through the linear psychological time of "past-present-future". That's why we see changes are running in some linear time. The conscious observer is aware of linear time; he experiences changes directly as they run in timeless space without the interference of psychological time. Symmetry in time is wrong imagination based on our experience of material changes via psychological time. With our elementary perception, we can only confirm material changes run in space.

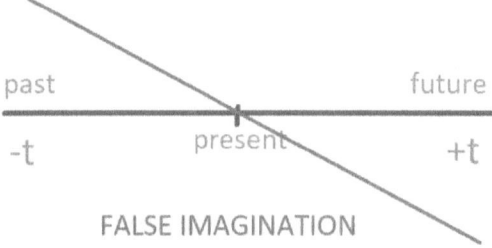

FALSE IMAGINATION

Time symmetry imagination is a concept that does not exist in physical reality. Considering time physical reality in which the universe run is somehow the old idea of 20th-century science which we should not carry on in 21 century. Material changes run in space only and time is their numerical sequential order. This 21 century understanding of time is the basis for cosmology and physics progress.

There is no physical Past, no physical Future, entire Universe is entangled into NOW

Time is what we measure with clocks. With clocks we measure numerical sequential order of events which run in space. Fundamental unit of the numerical order is Planck time Tp. Every elapsed time of a given event is the sum of Planck times.

$$T = Tp1 + Tp2 + Tp3,...,+Tpn.$$

Events which run in space are irreversible. Every event can be seen as the slices of Planck times. When slice Tp1 is in existence, the slice Tp2 is not in existence yet. When slice Tp2 enters existence, the slice Tp1 does not exist any more and the slice Tp3 is not existent yet.

Time is not 4th dimension of space. We have to drop this idea for which we do not have a single scientific evidence. Change run in space only and time is their numerical sequential order. Change do not run in space. This view is false and should be abandoned if we want progress physics. Theoretically seen if we do not have change in space there is no time. **Time is just the epiphenomena of change**. We are giving time too much importance. We believe that some physical past and future exists, that existence is symmetric in time. This view is false we do not have a single data it is right.

Time is the most controversial subject in physics. The case is nobody really understood what time really is. Einstein was fully aware that time has merely mathematical existence but he was not able to convey this to the scientific community. It is time we fix our understanding of time in physics. Without this there is no progress. The facts about time are following:

1. Universal space is time-invariant. Time when measured is merely the duration of change in space.

2. There is no physical future and no physical past. Changes run in space only. We experience timelessness of space (in the sense time is not its 4th physical dimension) as NOW.

3. In the universal space is always and only NOW.

4. The entire universe is entangled via universal space. Universal space is a medium in which transfer of information is immediate.

space	information transfer is imediate
photon	information transfer has light speed
airplane	information transfer is 900km/h
train	information transfer is 150km/h
walking postman	information transfer is 5km/h

Information transfer in the universe

Hundreds of articles was written on the subject of time travel which is pure illusion. We live in NOW which is the only physical reality. Linear time "past-present-future" exists only in the human mind. Einstein was aware of this 100 years ago, it is time now we enter the timeless nature of the universe.

14.   How big is the Weight of Life?

Back in 1987, I was a student at the University of Ljubljana, Slovenija. Once in a student campus, we have a big discussion of the origin of life, cosmology, and physics. I ask a friend who has studied physics if life is related to gravity. He said that life has nothing to do with gravity. The other friend who was biologist had the same opinion. I was a stubborn guy and I decided to measure eventual relation between life and gravity.

A group of professors gave me permission to use the most precise balances we had on our university. Between 1987 -90 I was measuring the difference of the weight between living and the same dead earthworms. Surprisingly the living mass had more weight than the same dead mass. I was measuring worms weight, alive one and the same dead one in a mass closed system. Dead one weight was diminishing for the one million part of the alive weight.

We can express this with the following equation:

$$F_g alive = F_g dead + \Delta F_g$$

Gravity works stronger on living mass than on the same dead mass. I published a few articles on the result of this experiment. Recently in NeuroQuantology in an article titled "Unified Field Theory Based on Bijective Methodology" on November 2018.

Last year I repeated the experiment again on **Mettler-Toledo mass comparator AX107H** which is 100-times more precise than the balance that I used in 1987–90. I got the same result as in my student years. I will do a few more repetitions and publish results hopefully in an established journal. Today science does not like discoveries which are our of the established paradigm. You can only discover new things in the frame of established science, which in physics means Standard model, in cosmology Big Bang model.

Mettler-Toledo mass comparator AX107H with 2 test tubes with distilled water and two test tubes with worms. Experiment was carried out in 2018.

You imagine that we could put a few earthworms in a deep freezer. They would fie immediately but their atomic structure would remain the same which means mass would remain the same. We can express this with the following formula:

$$m_{alive} = m_{dead}$$

Gravity acts stronger on the living mass than on the same dead mass. How that? Has a living organism special energy which the same dead organism does not have? Can this energy minimally increase the weight of a living organism? The answer is YES, in the living organism is a special kind of the energy which minimally increases organism weight. I was not the first one measuring this phenomenon. The first one was American physician Duncan MacDougall. He measured dying people and the difference was about 21 grams.

Dr. Duncan MacDougall

The **21 grams experiment** refers to a scientific study published in 1907 by Duncan MacDougall, a physician from Haverhill, Massachusetts. MacDougall hypothesized that souls have physical weight, and attempted to measure the mass lost by a human when the soul departed the body. MacDougall attempted to measure the mass change of six patients at the moment of death. One of the six subjects lost three-fourths of an ounce (21.3 grams).

In Dynamic Vacuum Relativity (see my book Relativity Reborn) which I developed in the last 5 years vacuum is multidimensional. Atoms are 3dimensional and exist in a 4dimensional layer of the vacuum. Subatomic particles are 4dimensional structures of the vacuum. More dimensional structures of the vacuum are also existing. These structures have no entropy and are actively involved in the functioning of the living organism. Their presence increases minimally the weight of the living organism.

Health is in the higher dimensional layers of the vacuum. In India, they called it PRANA in China QI. Our western science did not reach so far yet. They will one day. Better to come later than never. Western science is extremely pretentious in his idea that what can be measured is real, and what cannot be measured is not real. We are living in the illusion that what is "scientific" is real and what science cannot grasp is "nonscientific" and so unreal. We can measure only a small part of existence. Most of the existence you cannot measure, you can only know it by the experience. And human experience reaches far beyond the rational mind. Universe is a miracle and life is miracle. Step out of your "mind-box", step out of your "time-box" and the mystery will unveil.

Dear reader, I hope you like the book. Share it with your friends the messages that there is no God beyond the universe, the universe itself is God. We are living in a perfect universe and to see and experience thre universe directly we have to step out of the mind into the vastness of consciousness.

Scientific literature on the Renaissance of physics

1. Fiscaletti, D., Sorli, A. Perspectives of the Numerical Order of Material Changes in Timeless Approaches in Physics. *Found Phys* **45,** 105–133 (2015). https://doi.org/10.1007/s10701-014-9840-y
2. Fiscaletti, D., Sorli, A. Searching for an adequate relation between time and entanglement. *Quantum Stud.: Math. Found.* **4,** 357–374 (2017). https://doi.org/10.1007/s40509-017-0110-5
3. Šorli, A.S. Mass–Energy Equivalence Extension onto a Superfluid Quantum Vacuum. *Sci Rep* **9,** 11737 (2019). https://doi.org/10.1038/s41598-019-48018-2
4. Sorli, A. S. (2020). Black Holes are Rejuvenating Systems of the Universe . *JOURNAL OF ADVANCES IN PHYSICS*, *17*, 23-31. https://doi.org/10.24297/jap.v17i.8620
5. Sorli, A. S., & Čelan, Štefan. (2020). Integration of Life and Consciousness into Cosmology. *JOURNAL OF*

ADVANCES IN PHYSICS, 17, 41-49.
https://doi.org/10.24297/jap.v17i.8623

6. Sorli, A.S. (2020), Einstein's Vision of Time and Infinite Universe without Singularities - The End of Big Bang Cosmology, *JOURNAL OF ADVANCES IN PHYSICS* (accepted in publication)
7. Sorli, A.S. (2020) System Theory, Proton Stability, Double-Slit Experiment, and Cyclotron Physics, *JOURNAL OF ADVANCES IN PHYSICS* (accepted in publication)
8. Sorli, A.S.,Relativity Reborn – Bijective Physics (2019) Amazon (for experts)
9. Sorli, A.S. Intuitive Intelligence in Physics – Breaking the 10 Commandments (2020) Amazon (for nonexperts)